SPOTLIGHT ON ECOLOGY AND LIFE SCIENCE

HABITATS

SEBASTIAN AVERY

PowerKiDS press.

NEW YORK

Published in 2017 by The Rosen Publishing Group, Inc.
29 East 21st Street, New York, NY 10010

Editor: Caitie McAneney
Book Design: Michael Flynn
Interior Layout: Tanya Dellaccio

Photo Credits: Cover Brian Kinney/Shutterstock.com; p. 5 (Yellowstone National Park) Marc Turcan/Shutterstock.com; p. 5 (John Muir) https://commons.wikimedia.org/wiki/File:John_Muir_c1902.jpg; p. 7 Leeloona/Shutterstock.com; p. 8 John Cancalosi/Getty Images; p. 9 Matt Jeppson/Shutterstock.com; p. 10 David Osborn/Shutterstock.com; p. 11 (lion) Mark C. Ross/Getty Images; p. 11 (zebra) Phattana Stock/Shutterstock.com; p. 12 Planet Observer/UIG/Getty Images; p. 13 Mikadun/Shutterstock.com; p. 14 Ondrej Prosicky/Shutterstock.com;p. 15 Erni/Shutterstock.com; p. 17 (alligator) Renaud Visage/Shutterstock.com; p. 17 (bullfrog) Ilias Strachinis/Shutterstock.com; p. 17 (beaver) Christian Musat/Shutterstock.com; p. 18 Tosaphon C/Shutterstock.com; p. 19 Tom Murphy/Getty Images; p. 20 Ralph Lee Hopkins/Lonely Planet Images/Getty Images; p. 21 Derek Heasley/Shutterstock.com.

Cataloging-In-Publication Data

Names: Avery, Sebastian.
Title: Habitats / Sebastian Avery.
Description: New York : PowerKids Press, 2017. | Series: Spotlight on earth science | Includes index.
Identifiers: ISBN 9781499425642 (pbk.) | ISBN 9781499425673 (library bound) | ISBN 9781499425659 (6 pack)
Subjects: LCSH: Habitat (Ecology)--Juvenile literature. | Ecology--Juvenile literature.
Classification: LCC QH541.14 A94 2017 | DDC 577--d23

Manufactured in the United States of America

CPSIA Compliance Information: Batch #BW17PK For further information contact Rosen Publishing, New York, New York at 1-800-237-9932.

CONTENTS

CHAPTER ONE

AT HOME IN THE HABITAT

What does your home look like? What kinds of things do you have to protect you from the weather and conditions outside? You might use air conditioning if it's hot outside, or a heating system if it's cold outside. Animals have homes just like humans. Those homes are the animals' habitats.

There are many different kinds of habitats on Earth. Desert habitats are dry, while rain forests are wet. Over time, the animals that live in these environments have adapted, or changed to better survive there. Animals that live in cold places might have thicker fur to stay warm, while animals that live in hot places know how to beat the heat.

Habitats have everything an animal needs to survive, such as food, water, and places to live and hide. Read on to learn about different kinds of habitats and the animals that call them home.

JOHN MUIR

It's very important to keep natural habitats safe. John Muir was one of the most important nature lovers and activists who ever lived. He helped preserve habitats in California's Sierra Nevada Mountains and Yosemite Valley.

CHAPTER TWO

HIGH MOUNTAINS

Mountains are areas of high **elevation**. Some mountains stand alone, while others are part of huge mountain ranges such as the Himalayas. Temperatures get colder the higher you go. Certain trees and plants are found on different parts of mountains. The animals that live on mountains have often adapted to walking on steep, uneven land.

Snow leopards live in the mountains of central Asia. It's cold and dry in their high mountain habitat. The rugged land helps them hide and sneak up on their **prey**. Snow leopards can jump as far as 50 feet (15.2 m) to catch prey! Their grayish, white, and black fur helps them blend in with their rocky and sometimes snowy surroundings.

Rocky Mountain bighorn sheep make their home in mountains from Canada to New Mexico. They **developed** rough hooves that are split down the middle to help them balance on the rugged land.

Snow leopards' feet are wide and covered in thick fur. They act as snowshoes so the leopard can run and jump easily in its high mountain habitat.

CHAPTER THREE

DRY DESERTS

How do animals survive with little to no water? Some animals are perfectly suited to desert habitats. Deserts are places that get very little rainfall each year. Deserts are often hot, so the rain that does fall **evaporates** quickly. Many desert habitats have few plants that can survive the dry weather.

The kangaroo rat lives in deserts in the western United States. It doesn't mind the hot, dry habitat. In fact, it can go its whole life without drinking water!

SHORT-HORNED LIZARD

Reptiles often live well in deserts because they're cold-blooded. Cold-blooded animals have a body temperature that changes with the outside temperature. Lizards such as the short-horned lizard—also called the horned toad—lie in the hot desert sun to warm up and then find shade to cool down. The desert is the perfect place for the short-horned lizard to hide. Its tan, spiky skin fits in with the rocky, dry desert. When this lizard sees a predator, it can inflate its body to look like a spiky balloon or even shoot blood from its eyes!

CHAPTER FOUR

GRASSY PLAINS

Grasslands go by many names around the world—prairies, savannas, steppes, and rangelands. Grasslands are mostly flat lands that are covered in grasses and similar low-growing plants. Water is often in short supply in this habitat, but there's plenty of grass for **grazing**.

Bison once spread out across North American grasslands in great numbers. They roamed the land, eating grass as they went. As the biggest animal in North America, they need a lot of grass to survive. When Europeans arrived in North America, they killed bison until the animals nearly died out.

Zebras live in the savannas of eastern Africa. Like bison, they love to eat grass, as well as leaves, bark, and shrubs. They **migrate** often to find grassier land for grazing. Their stripes help them hide in high grass and confuse predators when they run in a group.

AMERICAN BISON

Because many plant-eating animals, such as zebras and antelopes, live in the savanna, predators, such as lions, like to hunt here.

CHAPTER FIVE

RAIN FOREST HOMES

Rain forests are lush, green habitats. The air is humid, or thick with water. The many trees in the forest let water vapor out into the air. This is called **transpiration**. It helps form thick clouds that hang above the trees. Rain forests are in tropical regions, which are warm and wet. Many animals live in the trees their whole life, never even touching the ground!

These side-by-side images show the effects of deforestation, or cutting down many trees, in the Amazon rain forest.

Jaguars' **adaptations** help them hunt both on land and in water.

Spider monkeys make their home in the rain forest habitats of Central and South America. They have long arms that help them swing from tree limb to tree limb. Their tails can grip branches too. They find most of their food, including fruits, leaves, eggs, and nuts, in the trees.

Jaguars are one of the great predators of the rain forest. Many live in the Amazon basin. With large, powerful bodies, jaguars are masters of their habitat and can easily climb trees and swim in rivers.

CHAPTER SIX

POLAR CHILL

The land around the North and South Poles is an **extreme** habitat. Antarctica in the south and the Arctic Circle in the north are the coldest places on Earth. Temperatures in the winter dip far below zero. The sun doesn't come out for months at a time. How can animals live in this kind of habitat?

Polar bears are the kings of the Arctic **ecosystem**. They have nearly **transparent** fur, which covers black skin. Their skin absorbs heat from whatever sunlight is available. Their big, furry paws help them walk on ice. They also have a thick layer of fat to keep them warm.

Arctic foxes have fur that changes color with the seasons. In the winter, their fur is white to match the snow. In the spring, their fur turns gray or brown to match the rocky land.

POLAR BEAR

Artic foxes will eat almost anything, which helps them survive. They'll even eat a polar bear's leftovers!

CHAPTER SEVEN

CREATURES OF THE MARSH

Marshes are wetlands, or places where water often covers the ground. They're found near rivers and lakes. Some marshes have freshwater, while others are made salty by ocean tides. The muddy bottom of a marsh is full of **nutrients**, which feed long grasses and other plants.

Beavers are just one of the many creatures that call the marsh home. They build homes in the middle of wetlands using trees, branches, and mud. They create an underwater entrance to keep predators out. Beavers have webbed feet and a flat tail that's shaped like a paddle. That makes them great swimmers.

American bullfrogs thrive in marsh habitats. They have webbed feet for swimming, long legs for jumping, and a huge mouth for catching food. They are green or brown and blend in with the marsh plants. Bullfrogs love to eat all the bugs and fish a marsh offers.

The Everglades is the largest freshwater marsh in the United States. The American alligator lives in this habitat.

CHAPTER EIGHT

FRESHWATER HABITATS

Rivers, lakes, and ponds usually have freshwater. That's water without much salt in it. More than 100,000 species live in freshwater ecosystems. They're home to many kinds of fish. Many land animals live near bodies of freshwater because the water is good to drink.

The Mekong giant catfish is the largest freshwater fish without scales. These fish live in the Mekong River in Southeast Asia. They can grow up to 10 feet (3 m) long!

Catfish live on the bottoms of lakes and rivers around the world. They have body parts called barbels that look like whiskers. They use the barbels for tasting, smelling, and sensing what's in the water around them. That makes them great at hiding and hunting in murky water.

North American river otters often live on the edge of rivers. They build burrows with many tunnels that lead to the water. Their fur **repels** water, and their webbed feet and strong tail help them swim and hunt for long periods of time. They are very graceful swimmers.

CHAPTER NINE

UNDER THE SEA

Oceans cover nearly three-fourths of Earth's surface. We have only explored a small part of these huge bodies of salt water. Some animals live near the surface of the water or in shallow water, while others live miles below the surface in the deep sea. Each species had to adapt to its specific ocean habitat.

Coral reefs are habitats with a ridge of coral skeletons and living coral. Many coral reef fish have had to adapt to the shape of the reef. Some fish, such as the blue tang, have a flat body to help them swim in and out of narrow holes and hiding places in a reef.

The deep sea has near-freezing and pitch-black waters. Fish here need to make their own light. Lantern fish have body parts called photophores that glow to attract prey.

CORAL FOSSIL

Sharks are the top predators of the ocean. They're built for the hunt with razor-sharp teeth, amazing senses, and torpedo-shaped bodies.

CHAPTER TEN

THE IMPORTANCE OF HABITATS

Next time you see an animal, look at its features. Does it have thick fur? Does it have webbed feet? That might give you an idea of where it's meant to make its home.

It's important to keep natural habitats the way they are. Unfortunately, human activities such as **climate change** and cutting down trees have changed the habitats of many animals around the world. Because of this, they're at risk of dying out.

Scientists and engineers are working together every day to use new technology, such as **drones** and tracking devices, to learn more about habitats and how to keep them safe. What can you do to help? Educate people about the importance of natural habitats. When you visit habitats, such as the ocean or desert, don't disturb the animals or plants. With people's help, animals can stay at home in their habitats for many years to come.

GLOSSARY

adaptation (aa-dap-TAY-shun) A change in a type of animal that makes it better able to live in its surroundings.

climate change (KLY-muht CHANJ) Change in Earth's weather caused by human activity.

develop (dih-VEH-luhp) To change over time.

drone (DRONE) A pilotless aircraft.

ecosystem (EE-koh-sih-stuhm) A natural community of living and nonliving things.

elevation (eh-luh-VAY-shun) Height above sea level.

evaporate (ih-VAA-puh-rayt) To change from a liquid to a gas.

extreme (ehk-STREEM) Great or severe.

graze (GRAYZ) To eat grass.

migrate (MY-grayt) To move from one place to another.

nutrient (NOO-tree-ehnt) Something taken in by a plant or animal that helps it grow and stay healthy.

prey (PRAY) An animal hunted by other animals for food.

repel (reh-PEHL) To keep away.

transparent (trans-PAA-rent) Allowing light through.

transpiration (trans-puh-RAY-shun) The passage of water vapor from a plant into the air.

INDEX

PRIMARY SOURCE LIST

Page 5
John Muir, shown seated on rock with lake and trees in background. Photograph. ca. 1902. Now kept in the Library of Congress, Washington, D.C.

Page 12
An overhead view of east Rondônia, Brazil, showing the effects of deforestation. Left image taken in 1986. Right image taken in 2001. Satellite images.

Page 20
Fossilized coral. Located in Aldabra Atoll, Aldabra Islands, Seychelles, East Africa.

WEBSITES

Due to the changing nature of Internet links, PowerKids Press has developed an online list of websites related to the subject of this book. This site is updated regularly. Please use this link to access the list: www.powerkidslinks.com/sels/habi